MEGA MACHINES!

SUPER SUBMARINES

BY NATALIE HUMPHREY

Gareth Stevens PUBLISHING

Please visit our website, www.garethstevens.com. For a free color catalog of all our high-quality books, call toll free 1-800-542-2595 or fax 1-877-542-2596.

Portions of this work were originally authored by Kenny Allen and published as *Submarines*. All new material in this edition is authored by Natalie Humphrey.

Cataloging-in-Publication Data

Names: Humphrey, Natalie.
Title: Super submarines / Natalie Humphrey.
Description: New York : Gareth Stevens Publishing, 2023. | Series: Mega machines! | Includes glossary and index.
Identifiers: ISBN 9781538283196 (pbk.) | ISBN 9781538283219 (library bound) | ISBN 9781538283226 (ebook)
Subjects: LCSH: Submarines (Ships)–Juvenile literature.
Classification: LCC VM365.H86 2023 | DDC 623.825'7–dc23

Published in 2023 by
Gareth Stevens Publishing
2455 Clinton Street
Buffalo, NY 14224

Designer: Deanna Paternostro
Editor: Natalie Humphrey

Photo credits: Cover, p. 1 noraismail/Shutterstock.com; pp. 3, 4, 6, 8, 10, 12, 14, 16, 18, 20, 21 (bottom), 22, 23, 24 Nataliia K/Shutterstock.com; p. 5 Aerial-motion/Shutterstock.com; p. 7 Michael Rosskothen/Shutterstock.com; p. 9 Fouad A. Saad/Shutterstock.com; p. 11 Roberto Galan/Shutterstock.com; p. 13 Juan Corona/Shutterstock.com; p. 15 Charles-Edouard Cote/Shutterstock.com; p. 17 UMB-O/Shutterstock.com; p. 19 Cristian Puscasu/Shutterstock.com; p. 21 (top) Kuleshov Oleg/Shutterstock.com.

Printed in the United States of America

CPSIA compliance information: Batch #CW23GS: For further information contact Gareth Stevens at 1-800-542-2595.

CONTENTS

Boldface words appear in the glossary.

Deep Sea Ships

Submarines are a special kind of mega ship. Submarines can sail on top of the ocean. But when the crew is ready, submarines dive deep into the water! Submarines can stay **submerged** for weeks. One submarine stayed underwater for 111 days in a row!

Sturdy Steel

Submarines are most often made from a strong metal called steel. They are shaped like a cylinder with two narrow ends. **Officers** control the submarine from the control room. In the control room, officers can choose when to dive or come back to the **surface**.

To Sink or Float

Submarines have tanks that help them to **sink** or float. When the tanks are filled with air, the submarine floats. When a submarine is ready to dive, the tanks fill with water. To come back up, the tanks fill up with air!

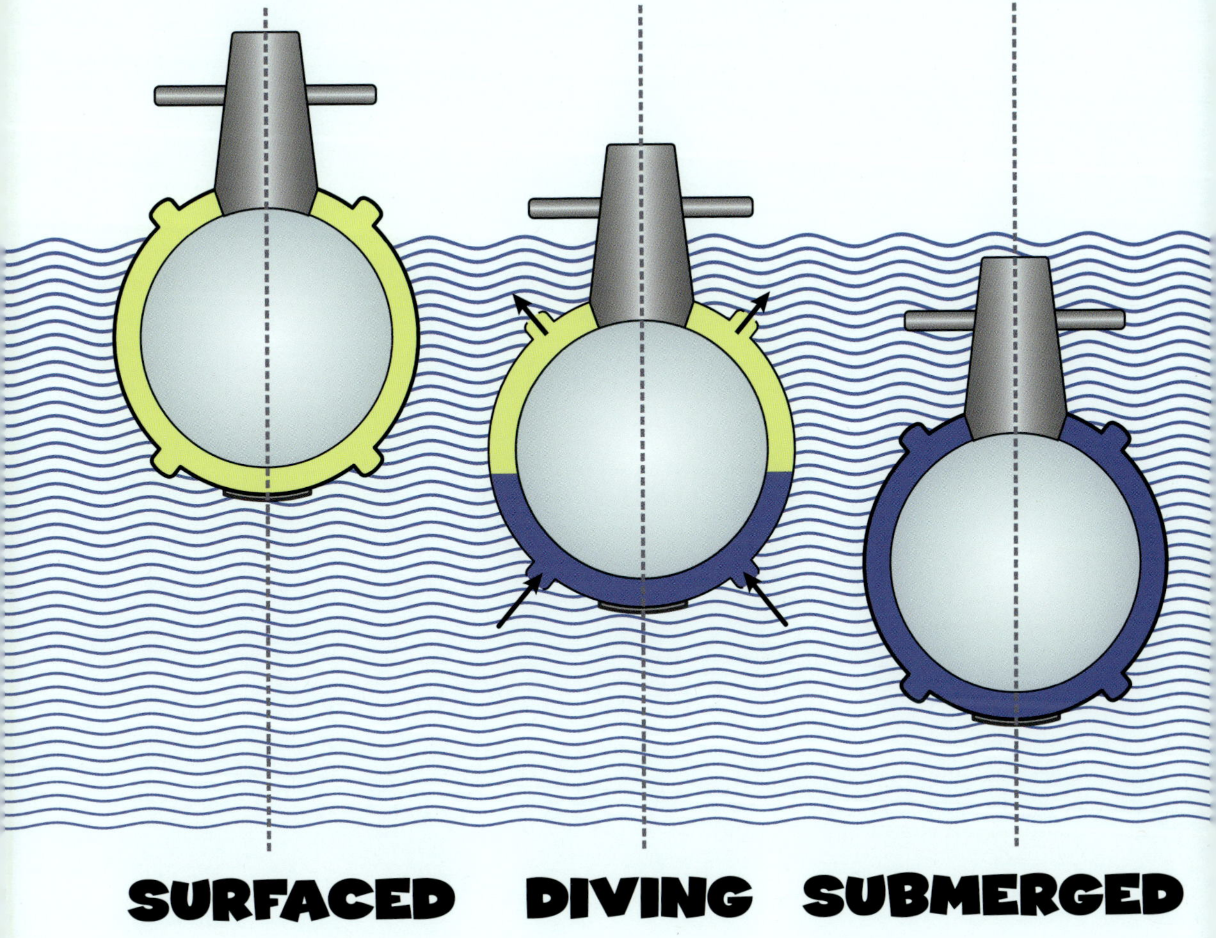
SURFACED
DIVING
SUBMERGED

The Crew

Submarine crews are made of navy **volunteers**. Depending on the class, or type, of submarine, crews will stay on board for different amounts of time. Some crew missions, or jobs, are a few weeks long. Longer missions can be around six months.

STAL CARO

Daily Life

Submarines usually carry small crews. Most often, crew members work for 8 hours then have 16 hours of free time. During their free time, crew members can eat, sleep, play cards, or watch TV.

Many submarines are so small, crew members share beds!

Powerful Engines

Submarines are powered by engines. Engines are machines that create power for the submarine to run. Some submarines have engines that burn fuel. Some have steam engines. Most **military** submarines today use **nuclear reactors**. Some submarines use more than one kind of engine!

Full Speed Ahead!

Submarine engines power propellers. Propellers have paddle-like parts that spin in the water to move the ship forward or backward. The fastest submarine ever built was Russia's K-222. At top speed, the K-222 could go 51 miles (82 km) per hour!

Different Jobs

Not all submarines are used by the military. Some submarines are used for undersea jobs. These submarines find lost ships or new kinds of fish and ocean animals. Some are used to repair oil platforms. Some submarines take people on underwater trips!

The Typhoon-Class

The Typhoon-class submarine is one huge mega machine! Created in Russia, the Typhoon-class is 575 feet (183 m) long. It can carry up to 175 crew members and can stay underwater for about 120 days. Only one Typhoon-class submarine is still in **commission**.

Mega Machine Facts: Typhoon-class submarine *Dmitry Donskoy*

Length: 600 feet (183 m)

Number of Crew Members: 150 to 175

Speed: 26 miles (42 km) per hour

Dive Depth: 1,300 feet (400 m)

GLOSSARY

commission: In active service of a military.

depth: How deep something is below the surface.

military: The armed forces.

nuclear reactor: A place where tiny particles called atoms are split to create energy.

officers: People in the military who are in charge of a military crew.

sink: Going below the surface of a liquid.

submerge: To put or sink below the surface of water.

surface: The top of a liquid.

volunteer: A person who offers a service without being asked.

FOR MORE INFORMATION

BOOKS

Rogers, Marie. *Incredible Submarines.* New York, NY: PowerKids Press, 2022.

Swanson, Jennifer. *How Do Submarines Work?* Mankato, MN: The Child's World, 2022.

WEBSITES

Kids Britannica: Submarine
kids.britannica.com/kids/article/submarine/390261
Check out more pictures and interesting facts about submarines.

Submarine Force Library and Museum Association
www.ussnautilus.org/
Learn more about submarines firsthand aboard the USS *Nautilus.*

Publisher's note to educators and parents: Our editors have carefully reviewed these websites to ensure that they are suitable for students. Many websites change frequently, however, and we cannot guarantee that a site's future contents will continue to meet our high standards of quality and educational value. Be advised that students should be closely supervised whenever they access the internet.

INDEX